Advanced Clinical Therapies in Cardiovascular Chinese Medicine

Advanced Clinical Therapies in Cardiovascular Chinese Medicine

Dr. Anika Niambi Al-Shura, BSc., MSOM, Ph.D
Continuing Education Instructor
Niambi Wellness
Tampa, FL

Medical Illustrator: Samar Sobhy

AMSTERDAM • BOSTON • HEIDELBERG • LONDON
NEW YORK • OXFORD • PARIS • SAN DIEGO
SAN FRANCISCO • SINGAPORE • SYDNEY • TOKYO
Academic Press is an imprint of Elsevier

Academic Press is an imprint of Elsevier
32 Jamestown Road, London NW1 7BY, UK
The Boulevard, Langford Lane, Kidlington, Oxford, OX5 1GB, UK
Radarweg 29, PO Box 211, 1000 AE Amsterdam, The Netherlands
225 Wyman Street, Waltham, MA 02451, USA
525 B Street, Suite 1900, San Diego, CA 92101-4495, USA

British Library Cataloguing-in-Publication Data
A catalogue record for this book is available from the British Library

Library of Congress Cataloging-in-Publication Data
A catalog record for this book is available from the Library of Congress

ISBN: 978-0-12-800122-6

For information on all Academic Press publications
visit our website at **store.elsevier.com**

This book has been manufactured using Print On Demand technology. Each copy is produced to order and is limited to black ink. The online version of this book will show color figures where appropriate.

DEDICATION

The energy and effort behind the research and writing of this textbook is dedicated to my son, Khaleel Shakeer Ryland. May this inspire and guide you through your journey in your medical studies, career, and life.

ACKNOWLEDGMENTS

This is a special acknowledgment to my 7-year medical students at Tianjin Medical University (2012–2013) who served as cardiovascular research assistants. May your future medical careers be successful.

An Qi He
Bin Lin Da
Han Jiang
Chen Hua
Jia Ying Luo
Jun Zhang
Lin Lin
Ming Lu
Nang Zhang
Ping Tang
Hu Si Le
Zhao Tian Man
Wen Xing Ning
Xing Wen Zhao
Tang Ying Mei
Li Ying Ying
Xiong Yong Qin
Ding Yu
Li Yan Jun

CONTENTS

SECTION II CARDIAC RHYTHM DISEASES

SECTION III CHAMBER AND VALVE DISEASES

SECTION IV VASCULAR SYSTEM DISEASES

INTRODUCTION

The companion course for this textbook edition can be found on the Elsevier website and at www.niambiwellness.com.

APPROVING AGENCIES

The course is approved by the National Certification Commission for Acupuncture and Oriental Medicine (NCCAOM) for 9.5 PDA points and Florida Board of Acupuncture for 5 CEUs.

The course with this textbook is entitled, Integrative Cardiovascular Clinical Therapies.

This course is approved by the National Certification Commission for Acupuncture and Oriental Medicine (NCCAOM), and is listed as course #1053-003 for 9.5 PDA points.

This course is approved by the Florida State Board of Acupuncture, and is listed as course # 20-334317 for 10 CEU credits.

COURSE DESCRIPTION

This informational course describes cardiovascular treatments using Chinese medicine and Western medicine.

COURSE OBJECTIVES

- Participants can use the information as a reference for understanding the key points of diseases that affect each section of anatomy, and how they play an overall role in heart failure.
- Participants can analyze the basic mechanisms of cardiovascular drugs and formulas, indications and contraindications for integrating medicines and genetic factors involved in each cardiovascular disease as a reference.
- Participants can use the information to plan for specialization in private practice.
- Participants can use the information to know when to refer to a Western medicine specialist or for practicing in an interdisciplinary setting.

SECTION I

Overview of Diseases

CHAPTER 1

Cardiovascular Disorders in TCM

CHAPTER OBJECTIVES

After studying this chapter, you should be able to:

1. List the cardiovascular diseases which are commonly treated in the TCM clinic.
2. Describe the aims of treatment using TCM alone or integrating it with Western medicine.
3. Compare and contrast constitutional factors versus genetic determinants.

1.1 PART 1: CONSTITUTIONAL THEORY

Constitutional theory in Chinese medicine is explained as a measurement of the patient's ability to adapt well to dietary, emotional, lifestyle, and environmental factors. It can be a method of determining the magnitude and duration of disease state and at times a guide for personalized treatment and prognosis.

1.1.1 Constitutional Theory Versus Genetics Determinants

In Chinese medicine, the constitution can be a postnatal acquired state based on dietary, emotional, lifestyle, and environmental factors. Genetics is part of the innate endowment. The body constitution and genetics will help diagnose and determine treatment. However, genetics can determine the prognosis and quality of life.

1.2 PART 2: LIPOPROTEIN DISORDERS

Hepatic, intestinal functioning, and renal disorders are well-known primary factors related to lipid transport and accumulation. Chinese medicine relies on Western medicine methods to confirm diagnosing and direction for treating serious lipoprotein disorders.

Advanced Clinical Therapies in Cardiovascular Chinese Medicine. DOI: http://dx.doi.org/10.1016/B978-0-12-800122-6.00001-2

The main concern is to protect and monitor the arteries, against atherosclerosis and coronary artery disease.

1.3 PART 3: HYPERTENSION

Hypertension is considered in Western medicine as high arterial blood pressure read at >140 or >90 in a seated or resting individual or more than one visit. There are two types of hypertension including essential or idiopathic hypertension, which is believed to have no definitive cause.

In Chinese medicine, severe emotions, improper nutrition, and visceral yin and yang disharmonies are the main reasons for secondary complications in controlling hypertension.

1.4 PART 4: ARRHYTHMIA

Since several types of arrhythmia result in sudden death, a diagnosed patient often requires close monitoring using Western medicine methods.

Acute atrial fibrillation may be due to a recent illness, temporary emotional, or lifestyle factors. This arrhythmia type may be suitable for treatment in the integrative medicine setting.

1.5 PART 5: SICK SINUS SYNDROME

As a prevention strategy, genetic determinants can be detected early in life to avoid certain risk factors. In a young patient with acute arrhythmia, this can be a tool to direct the treatment and to recover the patient and attempt to avoid sick sinus syndrome development later in life.

1.6 PART 6: ANGINA PECTORIS

The four types of angina pectoris include stable, unstable, variant, and microvascular.

Treatment success using integrative medicine focuses mostly on prevention and control during the stable angina stage.

1.7 PART 7: RHEUMATIC FEVER

Rheumatic fever is an autoimmune inflammatory disease. Within 20 days after infection, symptoms of rheumatic fever can present and weaken the heart. The immune responses play a role in defense and also in the autoimmunity process.

Early prevention may include the use of herbal formulas. If a patient is diagnosed with strep throat, they should be monitored closely, and it may be indicated to used treatments in integrative medicine.

1.8 PART 8: ENDOCARDITIS

Endocarditis is a chronic complication of rheumatic fever. Early prevention, especially when the treatment course for when the patient was diagnosed with strep throat, may include the use of integrative medicine.

Special complications:

Special Endocarditis Complications	
Heart	-murmur, heart failure and valvular damage
Heart	-murmur, heart failure and valvular damage -nosocomial endocarditis: contraction due to infection of artificial pacemakers, implants and intravenous lines
Brain	emboli or meningitis
Lungs	pneumonia and emboli due to right heart endocarditis
Kidneys	glomerular nephritis due to a metastasized piece of vegetation
Spleen	glomerular nephritis due to a metastasized piece of vegetation

1.9 PART 9: CARDIOGENIC SHOCK

The combination of Chinese medicine and Western medicine treatment is for inpatient care in the hospital setting.

The integration of Chinese medicine and Western medicine is to recover a surviving patient who has experienced cardiogenic shock.

Signs of Cardiogenic Shock	
Systolic blood pressure	<80mmHg
Cardiac index	</=2.2L/min m2
Pulmonary capillary pressure	>/=15mm Hg
Left chamber end diastolic pressure	>18mm Hg
Right ventricular end diastolic pressure	>10-15mm Hg

1.10 PART 10: THROMBOANGIITIS OBLITERANS

The condition is thought to occur as a result of tobacco use. The disease is characterized by inflammation and thrombosis of small and medium arteries of the legs and feet which is recurring and progresses in stages.

As in Western medicine, treatments using Chinese medicine should be seen as a palliative care option which mostly addresses symptoms.

1.11 PART 11: TAKAYASU ARTERITIS

Takayasu arteritis is also called "pulseless disease" and primarily affects various arteries in females between the ages of 20–40. There are six types that affect various major arteries which can lead to stenosis and are associated with kidney and heart failure.

Treatment using Chinese medicine can work with Western medicine to relieve physical symptoms such as pain, weakness, and emotional disturbances.

NOTES

CHAPTER 2

Listing of Primary TCM Differentiations for Cardiovascular Disorders

CHAPTER OBJECTIVES

After studying this chapter, you should be able to:

1. Explain the common TCM patterns and differentiations between different Western medicine diseases.
2. Describe the possible similarities between cardiovascular diseases with similar differentiations.
3. Describe the possible differences between cardiovascular diseases with similar differentiations.

Generally in TCM, practitioners expect that the spectrum of differentiations could apply to any syndrome. However, when focusing on cardiology as a TCM specialty, patterns and presentations have specific differentiations which become basic standards to guide clinical practice.

2.1 PART 1: LIPOPROTEIN DISORDERS

Hyperlipidemia is considered a pathological deficiency between the liver, spleen, and kidneys.

1. ______________________________
2. ______________________________
3. ______________________________

Visit the course for the answers.

2.2 PART 2: HYPERTENSION

In Chinese medicine, hypertension is considered a direct injury to the heart, liver, and kidneys.

Advanced Clinical Therapies in Cardiovascular Chinese Medicine. DOI: http://dx.doi.org/10.1016/B978-0-12-800122-6.00002-4

4. ______________________________
5. ______________________________

   Visit the course for the answers.

6. ______________________________
7. ______________________________
8. ______________________________

   Visit the course for the answers.

2.3 PART 3: ATRIAL FIBRILLATION

Chinese medicine differentiates arrhythmia as blood stasis, qi and yin deficiency, blocking of the heart vessel, phlegm, yang deficiency, cold, and dampness.

2.4 PART 4: ARRHYTHMIA

Chinese medicine differentiates arrhythmia as:

9. ______________________________
10. ______________________________
11. ______________________________
12. ______________________________
13. ______________________________
14. ______________________________

    Visit the course for the answers.

2.5 PART 5: SICK SINUS SYNDROME

However, the patterns which sequence toward the end stages in sick sinus syndrome are as follows:

15. ______________________________
16. ______________________________
17. ______________________________
18. ______________________________
19. ______________________________

    Visit the course for the answers.

2.6 PART 6: ANGINA PECTORIS

Xiong bi or chest obstruction is the disease classification in Chinese medicine. The differentiations include:

20. ____________________________
21. ____________________________
22. ____________________________
    Visit the course for the answers.
23. ____________________________
24. ____________________________
    Visit the course for the answers.

2.7 PART 7: RHEUMATIC FEVER

In Chinese medicine, pathogenesis begins with a failure of ying qi and wei qi in protecting the body's defenses leading to lingering exogenous heat and damp evils.

25. ____________________________
26. ____________________________
    Visit the course for the answers.

2.8 PART 8: ENDOCARDITIS

Endocarditis is the result of untreated or undertreated infection during rheumatic fever. In Chinese medicine, it can be considered toxic heat. It originates due to a failure of ying qi and wei qi in protecting the body's defenses. Exogenous wind, damp, and heat evils transform into endogenous toxic evils which consume qi and blood and complicate disease within muscles, joints, skin, the meridians, and threaten the life force of the patient.

27. ____________________________
28. ____________________________
    Visit the course for the answers.

2.9 PART 9: CARDIOGENIC SHOCK

In Chinese medicine, shock belongs to jue and tuo of both yin and yang which means the complete exhaustion of yin and yang. Blood deficiency and depleted qi are factors to tonify qi and nourish yin first.

2.10 PART 10: TAKAYASU ARTERITIS

In Chinese medicine, Takayasu arteritis is classified under vessel bi syndrome and is related to complications of the disharmony between the liver, spleen, and kidneys. The condition is differentiated as toxic phlegm heat during the active phases of the disease and deficiency of qi and blood stasis which can transform into yang deficiency with yin excess during remission in the stable phase.

29. ______________________
30. ______________________
    Visit the course for the answers.
31. ______________________
32. ______________________
    Visit the course for the answers.

2.11 PART 11: THROMBOANGIITIS OBLITERANS

In Chinese medicine, TAO is called tuo ju and belongs to the category of gangrene. The pathogenic factors include qi and blood deficiency, blood stasis, cold accumulation, and toxic heat affection.

33. ______________________
34. ______________________
35. ______________________
    Visit the course for the answers.

NOTES

Module Review Questions

1. List the cardiovascular diseases which are commonly treated in the TCM clinic.
2. Describe the aims of treatment using TCM alone or integrating it with Western medicine.
3. Compare and contrast constitutional factors versus genetic determinants.
4. Explain the common TCM patterns and differentiations between different Western medicine diseases.
5. Describe the possible similarities between cardiovascular diseases with similar differentiations.
6. Describe the possible differences between cardiovascular diseases with similar differentiations.

Log on at www.niambiwellness.com to access the companion course and quiz for Module 1.

SECTION II

Cardiac Rhythm Diseases

CHAPTER 3

Key Points of Cardiac Rhythm Disorders

CHAPTER OBJECTIVES

After studying this chapter, you should be able to:

1. Explain certain Western medicine key points about arrhythmia.
2. Explain certain Western medicine key points about sick sinus syndrome.
3. Discuss the similarities.
4. Discuss the differences.

3.1 PART 1: ARRHYTHMIA

3.1.1 Key Points

- Arrhythmia may manifest as a result of inflammation, emotional, or lifestyle factors at any age, but incidence commonly increases with age.
- Neither always caused by genetic mutations nor is it familial.
- Genetic factors involving microRNA include valvular disease and atrial fibrillation.
- Three mutations on genes KCNE2, KCNJ2, and KCNQ1 involve the replacement of certain key amino acid proteins which make up the channels that regulate potassium flow.
- KCNQ1 gene may be triggered during menopause with diseases such as ovarian cancer, type II diabetes, and LQTS.
- Mechanisms of arrhythmia include:
  1. ______________________________
  2. ______________________________
  3. ______________________________

  Visit the course for the answers.
- Dilation of heart chambers not only causes blood pressure changes, they also activate RAAS which leads to remodeling and more fibrosis in the myocardium and the SA and AV nodes. Diseases such as
  4. ______________________________

Advanced Clinical Therapies in Cardiovascular Chinese Medicine. DOI: http://dx.doi.org/10.1016/B978-0-12-800122-6.00003-6

- II, III, aVF, and V5 are 5.________________ and the P waves are
  6. ________________________

  Visit the course for the answers.
- Chest X-ray: determines
  7. ____________________________
- Echocardiography: determines
  8. ____________________________

  Visit the course for the answers.

3.2 PART 2: SICK SINUS SYNDROME

3.2.1 Key Points

- The prevention window may likely be
  9. ____________________________
- Atrial fibrillation may manifest as a result of inflammation, emotional, or lifestyle factors at any age, but incidence commonly increases with age. Sick sinus syndrome may be an end-stage condition in arrhythmia.
- HCN channels are involved in the automaticity of the SA node, and mutations may be involved with sinus bradycardia.
- Atrial fibrosis is a complication in atrial fibrillation which may cause sick sinus syndrome.
- Elderly patients with sick sinus syndrome may exhibit
  10. ____________________________
- End stages may necessitate more palliative care.

NOTES

CHAPTER 4

Key Points of Cardiac Rhythm Disorders in TCM

CHAPTER OBJECTIVES

After studying this chapter, you should be able to:

1. Explain certain Chinese medicine key points about arrhythmia.
2. Explain certain Chinese medicine key points about sick sinus syndrome.
3. Discuss the similarities.
4. Discuss the differences.

4.1 PART 1: ARRHYTHMIA

4.1.1 Key Points

- Atrial fibrillation may manifest as a result of
  11. ______________________________
  12. ______________________________
  13. ______________________________ at any age, but incidence commonly increases with age.
- Atrial fibrillation is a form of arrhythmia which may be indicated for treatment in Chinese medicine.
- Tachycardia is often differentiated as blood stasis and qi and yin deficiency.
- Bradycardia is usually differentiated as blood stasis which blocks the heart vessel, phlegm, yang deficiency, cold, and dampness.

4.2 PART 2: SICK SINUS SYNDROME

4.2.1 Key Points

- Sick sinus syndrome is differentiated into qi deficiency and blood stasis, deficiency of heart and spleen, heart yang deficiency, fire excess from yin deficiency, and deficiency of yin and yang.

Advanced Clinical Therapies in Cardiovascular Chinese Medicine. DOI: http://dx.doi.org/10.1016/B978-0-12-800122-6.00004-8

- Middle-aged patients should be concerned about arrhythmia especially if the patient has a history of 14________________. Prevention methods using integrative Chinese medicine may help 15. ______________________________.
- The differentiations in this section are for following and assessing disease progression and determining the differences between the indication of Chinese medicine and referral of the patient for the pacemaker.
- The prevention window may likely be 15. ______________________________
- During the prevention period, the treatment principle is to balance the heart and kidneys and tonify the spleen. This helps to nourish yin, move blood, and tonify yang and qi.
- End stages may necessitate more palliative care.

NOTES

CHAPTER 5

Integrative Treatments for Rhythm Disorders

CHAPTER OBJECTIVES

After studying this chapter, you should be able to:

1. List the pharmaceutical drugs and actions for each disorder.
2. List the basic herbal formulas and action for each disorder.
3. Discuss the treatment suggestions.

5.1 PART 1: ARRHYTHMIA

5.1.1 Pharmaceutical Drugs

Medicine to control heart rate	Acute heart rate control setting
Esmolol: Class I, evidence C	For patients without accessory pathway
Metroprolol: Class I, evidence C	
Propanolol: Class I, evidence C	
Diliatazem: Class I, evidence B	
Verapamil: Class I, evidence B	
Amiodarone: Class IIa, evidence C	For patients with accessory pathways
Digoxin: Class I, evidence B	For patients with heart failure without accessory pathway
Amiodarone: Class I, evidence C	

5.1.2 Chinese Medicine Formulas

These formulas should always be modified with other herbs depending on the diagnostic findings and treatment principles.

Advanced Clinical Therapies in Cardiovascular Chinese Medicine. DOI: http://dx.doi.org/10.1016/B978-0-12-800122-6.00005-X

Medicine	Action
An shen ning xin Formula	Soothes the spirit
Fu mai decoction	Qi and blood deficiency
Shen Fu decoction	Recovers after illness, warms yang
Si Ni decoction	Warms coldness
Wendan decoction	Regulates qi and transforms phlegm
Linggui Shugan decoction	Phlegm retention
Xuefu Zhuyu decoction	Blood stagnation and stasis
Taohong Siwu decoction	Invigorate blood
Zhi gancao decoction	Helps recover from illness, regulates heart functioning
Guipi decoction	Nourishes blood

5.2 PART 2: SICK SINUS SYNDROME

5.2.1 Pharmaceutical Drugs and Treatments

Medicine or device	Action
Pacemaker	Artificially adjusts the atrial heart rhythm.
Radio frequency ablation	Destroying/ablating the atrio-ventricular tissue around the AV node to control the rates of fibrillation in the atria from affecting ventricles.
Antiplatelet maedicine	Inhibits formation on of thrombus, especially to prevent the development of clotting around pacemakers and similar devices, and the development of clotting as a risk factor for stroke.
Beta blocker	Blocks the binding receptor on vascular smooth muscle cells that respond to epinephrine, the stress response.
Calcium channel blocker	Blocks the binding receptor on vascular smooth muscle cells that respond to epinephrine, the stress response.
Cardiac glycoside	Digoxin may be used for arrythmias and in conjunction with a pacemaker.

A. Which medications act on the vascular smooth muscle?
B. Which medications act to control heart rhythm?

5.2.2 Chinese Medicine Formulas

Medicine	Action
Gui zhi tang	Warms to body
Shen fu decoction	Recovers after illness, warms yang
Gui pi tang	Nourishes and tonifies blood
Zhi gan cao tang	Helps recover from illness, regulates heart functioning
You gui zhen zhu decoction	Tonifies kidney yin and calms the spirit

A. Which formulas act to recover the body from an illness?

5.3 PART 3: BASIC METHOD FOR INDICATING OR CONTRAINDICATING

Chinese medicine formulas for the treatment of certain arrhythmias can benefit more during the prevention stages, especially where the patient is already found to be constitutionally predisposed to development. Though Western medicine physicians are aware of the side effects when prescribing, the drugs are often considered to have benefits greater than the side effects.

- Use caution when prescribing or providing herbal formulas and nutritional supplement during a treatment course with pharmaceutical drugs. 16.____________________ Some of the pharmaceutical drugs prevent fat absorption in the intestines and assist in the rise of HDL levels. However, they can also block the absorption of
  17. ____________________________
  Visit the course for the answers.

Patients with sick sinus syndrome are usually elderly patients with symptoms characterized by alternating patterns of tachycardia and bradycardia. This condition puts the patient at risk for congestive heart failure, stroke, and sudden death. Most Western medicine physicians will automatically advise

18. ____________________________
  Visit the course for the answers.

- Herbal medicine may not be indicated
  19. ______________________________
- Patients in moderate condition and not
  20. ______________________________

  Visit the course for the answers.

NOTES

Module Review Questions

1. Explain certain Western medicine key points about arrhythmia.
2. Explain certain Western medicine key points about sick sinus syndrome.
3. Discuss the similarities.
4. Discuss the differences.
5. Explain certain Chinese medicine key points about arrhythmia.
6. Explain certain Chinese medicine key points about sick sinus syndrome.

7. Discuss the similarities.
8. Discuss the differences.
9. List the pharmaceutical drugs and actions for each disorder.
10. List the basic herbal formulas and action for each disorder.
11. Discuss the treatment suggestions.

Log on at www.niambiwellness.com to access the companion course and quiz for Module 3.

SECTION III

Chamber and Valve Diseases

CHAPTER 6

Key Points of Chamber and Valve Disorders

CHAPTER OBJECTIVES

After studying this chapter, you should be able to:

1. Explain certain Western medicine key points about hypertension.
2. Explain certain Western medicine key points about rheumatic fever.
3. Explain certain Western medicine key points about endocarditis.
4. Discuss the similarities.
5. Discuss the differences.

6.1 PART 1: HYPERTENSION

6.1.1 Key Points

- A genetic predisposition or family history of hypertension increases risk.
- Lifestyle contributors:
  1. ______________________________

  Visit the course for the answers.
- Prehypertension is an early warning to modify lifestyle.
- Preexisting conditions like thyroid disease and sleep apnea may be a cause and can be managed with modifications in lifestyle modification.
- In the United States, occurrence of hypertension remains higher in 2.________________ at any age; and in 3.________________ than in Caucasians.
- In China, all Han men as well as the other 56 ethnic minorities of men are five times as likely to develop and die of hypertension-related complications as women.
- The JNC7 classification refers to an adult who is not considered acute or within an emergency hypertensive status and untreated with medication or dietary restrictions (see Chapter 9).
- From the normal range, each increment of 4.______________ mmHg increases cardiovascular disease risk two and half times.

Advanced Clinical Therapies in Cardiovascular Chinese Medicine. DOI: http://dx.doi.org/10.1016/B978-0-12-800122-6.00006-1

- Patients with cardiovascular, diabetes, or renal diseases are under different criteria: the target should be 5.____________ mmHg.
- Systolic elevation rather than diastolic is a stronger risk factor in individuals over 50.
- Treatment of hypertension requires a close relationship between physician and patient with high patient interest and compliance.

6.2 PART 2: RHEUMATIC FEVER

6.2.1 Key Points

- Rheumatic fever is an autoimmune inflammatory disease which is connected with type II hypersensitivity, humoral and cellular immune-mediated immune injury.
- M proteins are similar to the antigen located on myocardial cells and therefore such antibodies may attack cardiac myocytes.
- The 6.__________ Criterion is used to assist in diagnosing.
- 7.______________ is a condition in affected children with temporary poor motor skills, emotional changes, and contracture of the fingers.
- Carditis can be heard through auscultation and may mark 8. ________________________________
- Polyarthritis includes migratory joint pain.
- Limbs: joint pain, subcutaneous nodules around the joints, and a body rash in some patients called erythema marginatum.
- Antistreptolysin O test is often used for detecting streptococcus, glomerular nephritis and determining whether joint pain is caused by rheumatic fever or rheumatoid arthritis.

6.3 PART 3: ENDOCARDITIS

6.3.1 Key Points

- HLA-DQA1, CFHR1, and CFH3 are genetic determinants involved with predispositions to autoimmune diseases and nephritic syndrome.
- Rheumatic fever has the possibility of transforming into endocarditis later in life for some patients.
- 9.______________ criteria is often used in the diagnosis of endocarditis, by determining if the patient presents with either two major criteria, one major, and three minor criteria or five minor criteria.
- "FROM JANE" a mnemonic for endocarditis: 10. ________________________________
- A positive Allen's test may indicate a circulation problem or Buerger's disease.

NOTES

CHAPTER 7

Key Points of Chamber and Valve Disorders in TCM

CHAPTER OBJECTIVES

After studying this chapter, you should be able to:

1. Explain certain Chinese medicine key points about hypertension.
2. Explain certain Chinese medicine key points about rheumatic fever.
3. Explain certain Chinese medicine key points about endocarditis.
4. Discuss the similarities.
5. Discuss the differences.

7.1 PART 1: HYPERTENSION

7.1.1 Key Points

- In Chinese medicine, hypertension is included in the categories of xuan yun (pronounced as shoo-an yoon) (vertigo) and tou tong (pronounced as toe-tung) (headache).
- Two or more medications in either Western, Chinese medicine or integrated medicine may be needed to control blood pressure in some patients. Chinese medicine is considered most effective up to stage 1. In late stage 2 or 11.__________ close monitoring and western medicine drugs may be indicated with diuretics especially with more serious disease complications.
- Certain Western and Chinese medications for 12.__________ when used without professional supervision can cause temporary elevations in blood pressure.

Elevators of blood pressure	
Pharmaceutical drugs	Herbs
Broncho dilators	Ma huang
Cardiotonics	Gan cao Lu cha

Advanced Clinical Therapies in Cardiovascular Chinese Medicine. DOI: http://dx.doi.org/10.1016/B978-0-12-800122-6.00007-3

7.2 PART 2: RHEUMATIC FEVER

7.2.1 Key Points

- Wei qi and ying qi are part of the immune system and function to protect the body.

7.3 PART 3: ENDOCARDITIS

7.3.1 Key Points

- Exogenous wind, damp, and heat evils transform into endogenous heat toxins.

NOTES

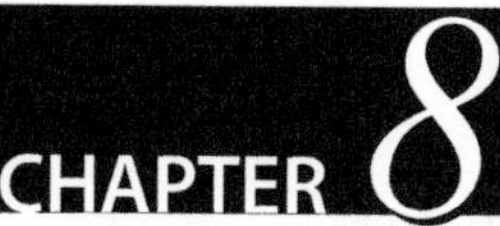

Integrative Treatments for Valve Disorders

CHAPTER OBJECTIVES

After studying this chapter, you should be able to:

1. List the pharmaceutical drugs and actions for each disorder.
2. List the basic herbal formulas and action for each disorder.
3. Discuss the treatment suggestions.

8.1 PART 1: HYPERTENSION

8.1.1 Pharmaceutical Drugs

Medicine	Action
Thiazide diuretics.	Reduces water
Beta blockers	Heart beats slower and with less force
ACE inhibitors.	Relaxes blood vessels
ARBs	Relaxes blood vessels
Calcium channel blockers	Relaxes blood vessels
Renin inhibitors.	Slows the production of renin
Alpha blockers	Reduces nerve impulses that narrow blood vessels
Alpha-beta blockers.	Reduces nerve impulses, slow the heartbeat
Central-acting agents	Prevents nervous system heart rate increase and blood vessel narrowing
Vasodilators.	Prevents the arteries from constriction

A. Which medications relax the blood vessels?
B. Which medications decrease heart beat?

Advanced Clinical Therapies in Cardiovascular Chinese Medicine. DOI: http://dx.doi.org/10.1016/B978-0-12-800122-6.00008-5

8.1.2 Chinese Medicine Formulas

Medicine	Action
Modified Gambirplant Branch Formula	Reduces hypertension, headache and dizziness
Modified Yang Hyperactivity Check with 7 Drugs Formula	Reduces hypertension, headache, tinnitus, and mildly diuretic
Modified Liver subduing and Wind Stopping Formula	Reduces hypertension, thyroid problems, hyper aldosterosim, & tranquilizes the mind
Modified Blood Pressure Reducing Decoction	Reduces blood pressure, dreaminess, insomnia, brain fog, chest oppression
Modified Qi Replenishing and Yin Nourishing Formula	Stablizes blood pressure while recovering from dryness, exhaustion and headache
Modified Cardiotonic Formula	Reduces blood pressure, nourishes blood
Diabetes, kidney failure, aldosteronism, renovascular disease, coarctation of aorta, thyroid disease or tumors	Reduces blood pressure in serious cases, and coordinates with pharmaceutical drugs

A. Which formulas reduce hypertension and headache?

8.2 PART 2: RHEUMATIC FEVER

8.2.1 Pharmaceutical Drugs

Medicine	Action
Aspirin	Anti-inflammatory, analgesic, antipyretic
Corticosteroids	Anti-inflammatory
Penicillin	Antibacterial
Erythromycin	Antibacterial
Azithromycin	Antibacterial

A. Which medications are anti-inflammatory?

8.2.2 Chinese Medicine Formulas

Medicine	Action
qing wen bai du yin	Treats severe toxic heat
pu ji xiao du yin	Treats severe toxic heat
long dan xie gan tang	Treats damp heat which affects the lower jiao
shen tong zh yu tang	Unblocks meridians and moves qi and blood

A. Which of the formulas has an action which might benefit a patient with an infection?
B. Which of the formulas has an action which might benefit a patient with pain?

8.3 PART 3: ENDOCARDITIS

8.3.1 Pharmaceutical Drugs

Medicine	Action
Intravenous vancomycin	Treats *Enterococcus* infection
Intravenous ceftrioxene	Treats resistant bacterial infections
Penicillin	Treats *Enterococci*, *Gemella* sp., *Granulicatella* sp. and *Abiotrophia defectiva*
Nafcillin	Treats *Staphylococcus* aureus
Aminoglycosides	Antibiotics which treat aerobic gram negative bacterial infections
Intravenous benzyl penicillin	Treats *Streptococcus* bovis

8.3.2 Chinese Medicine Formulas

Medicine	Action
Sheng yang yi wei tang	Resolves exterior pathogenic factors, drains dampness, raises yang
Ren shen tang	Replenishes qi, nourishes yin, raises yang, treats palpitations and vertigo
Qing wen bai du yin	*Clears toxic heat from blood, nourishes yin*
Pu ji xiao du yin	Clears toxic heat and wind damp-heat, reduces swelling
Long dan xie gan tang	Clears excess liver and gall bladder damp heat
Shen tong zhu yu tang	Activates blood circulation and qi flow, dredges meridians and collaterals and relieves pain

8.4 PART 4: BASIC METHOD FOR INDICATING OR CONTRAINDICATING

Chinese medicine formulas can benefit patients alone or integrated with pharmaceutical drugs for hypertension. It is best indicated for patients with mild to moderate hypertension. Patients in the moderate hypertension range where the blood pressure seems to fluctuate between 13.________________ may benefit more with combining Chinese and Western medicine.

Because endocarditis is a potentially life-threatening cardiovascular condition, Western medicine health providers may restrict treatment for endocarditis to antibiotic therapy. Some patients may benefit from integrating the Western medicine antibiotics with Chinese herbal antibiotics by substituted sulfa drugs in some patients with an allergy or low tolerance for side effects.

For rheumatic heart fever, some patients may benefit from integrating the Western medicine antibiotics with Chinese herbal antibiotics. Qing wen bai du yin may be modified and substituted for sulfa drugs in some patients with an allergy or low tolerance for side effects. In some cases it may be used alone. With either use in conjunction with antibiotic drugs, the overall condition will need to be monitored with laboratory tests.

NOTES

Module Review Questions

1. Explain certain Western medicine key points about hypertension.
2. Explain certain Western medicine key points about rheumatic fever.
3. Explain certain Western medicine key points about endocarditis.
4. Discuss the similarities.
5. Discuss the differences.
6. Explain certain Chinese medicine key points about hypertension.
7. Explain certain Chinese medicine key points about rheumatic fever.
8. Explain certain Chinese medicine key points about endocarditis.
9. Discuss the similarities.
10. Discuss the differences.
11. List the pharmaceutical drugs and actions for each disorder.
12. List the basic herbal formulas and action for each disorder.
13. Discuss the treatment suggestions.

Log on at www.niambiwellness.com to access the companion course and quiz for Module 6.

SECTION IV

Vascular System Diseases

CHAPTER 9

Key Points of Vascular Disorders

CHAPTER OBJECTIVES

After studying this chapter, you should be able to:

1. Explain certain Western medicine key points about hypertension.
2. Explain certain Western medicine key points about rheumatic fever.
3. Explain certain Western medicine key points about endocarditis.
4. Discuss the similarities.
5. Discuss the differences.

9.1 PART 1: HYPERTENSION

JNC7 Prevention, Detection, Evaluation, and Treatment of High Blood Pressure		
Classification	**Systolic mmHg**	**Diastolic mmHg**
Normal	<120	<80
Pre- Hypertension	120–139	80–89
Stage 1 Hypertension	140–159	90–99
Stage 2 Hypertension	>160	>100

- A genetic predisposition or family history of hypertension increases risk.
- Prolonged stress and anxiety related to environment can also increase risk.
- Prehypertension is an early warning to modify lifestyle.
- Preexisting conditions like 14.__________ may be a cause and can be managed with modifications in lifestyle modification.
- Lifestyle contributors:
  15. ____________________________

Advanced Clinical Therapies in Cardiovascular Chinese Medicine. DOI: http://dx.doi.org/10.1016/B978-0-12-800122-6.00009-7

- Birth control pill and hormonal replacement use.
- Current history of obesity.
- In the United States, occurrence of hypertension remains higher in adult males than females at any age; and in African Americans, Hispanics, and Asians than in Caucasians.
- Treatment of hypertension requires a close relationship between physician and patient with high patient interest and compliance.
- The JNC7 classification refers to an adult who is not considered acute or within an emergency hypertensive status and untreated with medication or dietary restrictions.
- From the normal range, each increment of 20/10 mmHg increases cardiovascular disease risk two and half times.
- Patients with cardiovascular, diabetes, or renal diseases are under different criteria: the target should be <130 or <80 mmHg.
- Systolic elevation rather than diastolic is a stronger risk factor in individuals over age 50.

9.2 PART 2: ANGINA PECTORIS

Signs include low oxygen distribution, blood clots, arterial spasms, and plaque deposits in the main and micro-arteries of the heart.

The four types of angina pectoris include:

Types of angina pectoris	
Stable angina	Regular episodes of chest pain which are relieved by rest and medication
Unstable angina	Irregular unpredietable episodes of chest pain not relieved by rest, but relieved by nitroglycerine
Microvascular angina	The micro arteries are blocked and cannot be detected on imaging; primarily occurs in women.
Variant angina	Occursat night and may lead to sudden death.

- Myocardial and Na^{+}/K^{+} pump failures increase H^{+} production which lowers pH.
- Troponin is a cardiac marker which elevates within a few hours after cardiac injury and persists for 16.________________.

- BNP/NT-proBNP is a cardiac marker, and levels indicate 17.________________.
- CK-MB (creatine kinase) is a cardiac marker which elevates when there is 18.________________.
- Coronary angiography and CABG: determines the magnitude of coronary artery block. Next healthy vessels from another area of the body are surgically used to bypass or replace the blocked section.

9.3 PART 3: HYPERLIPIDEMIA

Hyperlipidemia has at least three well known and two other factors related to the causes. Hepatic, intestinal functioning, and renal disorders are well-known primary factors related to lipid transport and accumulation, and metabolic and hormonal disorders are others.

- Hormonal causes of lipoprotein elevations include 19. ____________________________
- The metabolic causes include the symptoms found in metabolic syndrome and type II diabetes, both of which tend to elevate cholesterol levels. Symptoms include 20.________________.

Lipid type	mg/dL	mmol/L	Interpretation
Total cholesterol (TC)	<200 200–239 >240	<11.1 11.1–13.3 >13.3	Normal Borderline High
LDL	<100 100–129 130–159 160–189 >190	<5.5 5.55–7.15 7.15–8-82 8.82–10.5 >10.5	Normal Good Borderline High High risk
HDL	<40 41–59 60	<2.21 2.21–3.27 3.27	High risk Borderline Good

A. Which range is considered normal for LDL?
B. Which range is considered normal for HDL?

9.4 PART 4: THROMBOANGIITIS OBLITERANS

Chinese medicine or Western medicine treatment should be seen as

21. ____________________________

Stages of Thrombo angiitis obliterans	
Ischemic period	The veins of the lower leg are inflamed, skin has ulceration and gangrene with intermittent claudication
Nutritional disorder period	Persistent pain and muscle atrophy
Necrosis period	Pain becomes worse as infection leads to the development of ulcers and gangrene

9.5 PART 5: TAKAYASU ARTERITIS

Takayasu arteritis is also called "pulseless disease" and primarily affects females between the ages of 20–40. There are six types that affect various major arteries which can lead to stenosis and are associated with kidney and heart failure.

1990 American College of Rheumatology criteria	
Age of onset	< 40 years old
Claudication	Development and worsening of fatigue and discomfort in muscles of 1 or more extremity while in use, especially the upper extremities.
Brachial artery pulse	Decreased pulsation of 1 or both brachial arteries.
BP difference	Difference of >10 mm Hg in systolic blood pressure between arms.
Bruit	Bruit audible on auscultation over 1 or both subclavian arteries or abdominal aorta.
Arteriogram abnormality	Arteriographic narrowing or occlusion of the aorta, primary branches, or large arteries in the proximal upper or lower extremities. This is not due to atherosclerosis or fibro-muscular dysplasia.

A. What is the criteria for vessel problems?
B. What is the criteria for evidence of a bruit?

1986 Ishikawa criteria	
Age of onset	< 40 years of age with at least 1 month of symptom duration.
Major criteria	
Angiographic findings include lesions in the left and right mid sub-clavian artery. The most severe stenosis or occlusion presents in the mid portion of the artery. The location is from a 1 cm point which is proximal to the left and right of the vertebral artery orifices to a 3-cm distal point to the orifice.	
Minor criteria	
Angiographic and echocardiographic findings include annulo-aortic ectasia or aortic regurgitation. Also there may be lesions found in the pulmonary artery, left mid–common carotid, distal brachio-cephalic trunk, descending aorta, or abdominal aorta.	

A. What are the similarities between the major and minor criteria?
B. What are the differences between the major and minor criteria?

- It is considered an autoimmune disorder which primarily affects the aorta of females between age 20 and 40.
- Takayasu arteritis is a 22.________________ disease which affects large arteries and is characterized by the absence of a pulse, thus also called "pulseless disease."
- There are six types which affect sections of the aorta, abdominal, and renal arteries.
- The disease mechanisms include heredity, estrogen excess, and infection.
- Three factors from the 1990 American College of Rheumatology criteria can be used to diagnose.
- Two major and one minor criteria from the 1986 Ishikawa criteria can be used for diagnosis.

NOTES

CHAPTER 10

Key Points of Vascular Disorders in TCM

CHAPTER OBJECTIVES

After studying this chapter, you should be able to:

1. Explain certain Western medicine key points about hypertension.
2. Explain certain Western medicine key points about rheumatic fever.
3. Explain certain Western medicine key points about endocarditis.
4. Discuss the similarities.
5. Discuss the differences.

10.1 PART 1: HYPERTENSION

In Chinese medicine, hypertension is considered a direct injury to the heart, liver, and kidneys. It is a cyclic condition marked initially by imbalances between yin and yang due to prolonged emotional crisis, improper diet, and other factors resulting in the deficiency of both.

According to both Chinese and Western medicine, blood pressure is affected by psychoemotional responses, diet and digestion, lifestyle choices, and exercise practices. In addition, the time of day, season of the year, environmental pollution, and microbes are also contributing factors.

In Chinese medicine, severe emotions, improper nutrition, and visceral yin and yang disharmonies mainly lead to the reasons for complications in controlling hypertension.

- In Chinese medicine, hypertension is included in the categories of xuan yun (pronounced as shoo-an yoon) (vertigo) and tou tong (spelled as toe-tung) (headache).

Advanced Clinical Therapies in Cardiovascular Chinese Medicine. DOI: http://dx.doi.org/10.1016/B978-0-12-800122-6.00010-3

- Certain Western and Chinese medications like bronchodilators and ma huang (ephedra), cardiotonics and formula harmonizers like gan cao and antioxidants such as 100% lu cha (green tea) when used without professional supervision can cause temporary elevations in blood pressure.
- Lifestyle contributors: lack or insufficient physical activity, excessive smoking, excessive alcohol consumption, excessive sodium intake, insufficient balancing nutrients, and other minerals.
- Current history of obesity.
- In the United States, occurrence of hypertension remains higher in adult males than females at any age; and in African Americans, Hispanics, and Asians than in Caucasians.
- In China, all Han men as well as the other 56 ethnic minorities of men are five times as likely to develop and die of hypertension-related complications as women.
- Treatment of hypertension requires a close relationship between physician and patient with high patient interest and compliance.
- Two or more medications either in Western, Chinese medicine or integrated may be needed to control blood pressure in some patients. Chinese medicine is considered most effective up to stage 1. In late stage 2 or critical events >160/100 close monitoring and western medicine drugs may be indicated with diuretics especially with more serious disease complications.

10.2 PART 2: ANGINA PECTORIS

Xiong bi or chest obstruction is the disease classification in Chinese medicine. Treatment success using integrative medicine focuses mostly on prevention and control during the stable angina stage.

10.2.1 Seasonal Factors

Seasonal Factors	
Cold weather in winter	Patients experience deficiency of qi and yang, which causes stress during physical effort.
Hot weather in summer	Patients experience loss of qi and yin fluid, which raises body temperature and causes sweating.

10.2.2 Lifestyle Choices

- Angina pectoris is classified under xiong bi.

10.3 PART 3: HYPERLIPIDEMIA

Chinese medicine relies on Western medicine concerning the diagnosing methods and direction for treating lipoprotein disorders.

- In Chinese medicine, lipoprotein disorders are categorized as tanzhuo (pronounced as tan-joo-oh), xueyu (pronounced as shoo-eh-yoo), and xuan yuan (pronounced as shoo-an-yoon) and are classified under 23.______________
- Because it is a factor in other diseases, it may also belong to the categories of headache, stroke, palpitation, obstruction, and vertigo.
- Like hypertension, hyperlipidemia is more a symptom rather than a disease itself. In Chinese medicine, hyperlipidemia is considered a pathological deficiency between the liver, spleen, and kidneys.
- With depletion of liver yin, hyperactive liver yang overacts upon the spleen which results in an acute digestive impairment.
- Prolonged excess liver yang overaction results in chronic digestive problems.

10.4 PART 4: THROMBOANGIITIS OBLITERANS

In Chinese medicine, TAO is called tuo ju with five characteristics. It is also known as vasculitis and belongs to the category of gangrene. The pathogenic factors include qi and blood deficiency, blood stasis, cold accumulation, and toxic heat affection.

As in Western medicine, treatments using Chinese medicine should be seen as a palliative care option which mostly addresses symptoms. Treatment courses are often effective, however frequent relapse and amputation is the realistic prognosis.

Five Characteristic of Tuo Ju
Congenital deficiencies
Weak kidney qi, yin and yang
Cold and dampness
Abnormal diet
Poor psycho-emotional response

10.5 PART 5: TAKAYASU ARTERITIS

In Chinese medicine, takayasu arteritis is classified under vessel bi syndrome and is related to complications of the disharmony between the liver, spleen, and kidneys. The condition is differentiated as toxic phlegm heat during the active phases of the disease and deficiency of qi and blood stasis which can transform into yang deficiency with yin excess during remission in the stable phase.

- It is considered an autoimmune disorder which primarily affects the aorta of females between age 20 and 40.
- Takayasu arteritis is a 24.________________ disease which affects large arteries and is characterized by the absence of a pulse, thus also called "pulseless disease."
- In Chinese medicine, the disease is divided into 25. ________________ phases.

NOTES

Log on at www.niambiwellness.com to access the companion course and quiz for Module 6.

CHAPTER 11

Integrative Treatments for Vascular Disorders

CHAPTER OBJECTIVES

After studying this chapter, you should be able to:

1. List the pharmaceutical drugs and actions for each disorder.
2. List the basic herbal formulas and action for each disorder.
3. Discuss the treatment suggestions.

11.1 PART 1: HYPERTENSION

11.1.1 Pharmaceutical Drugs

Drug	Action
Thiazide diuretics	Reduces edema
Beta blockers	Heart beats slower and with less force
ACE inhibitors	Relaxes blood vessels
ARBs	Relaxes blood vessels
Calcium channel blockers	Relaxes blood vessels
Renin inhibitors	Slows the production of renin
Alpha blockers	Reduces nerve impulses, slow the heart beat
Central-acting agents	Prevents heart rate increase and blood vessel narrowing
Vasodilators	Prevents the arteries from constriction
Thiazide	Diuretics reduces edema

Advanced Clinical Therapies in Cardiovascular Chinese Medicine. DOI: http://dx.doi.org/10.1016/B978-0-12-800122-6.00011-5

11.1.2 Chinese Medicine Formulas

Medicine	Action
Modified Gambirplant Branch Formula	Reduces hypertension, headache and dizziness
Modified Yang Hyperactivity Check with 7 Drugs Formula	Reduces hypertension, headache, tinnitus, and mildly diuretic
Modified Liver subduing and Wind Stopping Formula	Reduces hypertension, thyroid problems, hyper aldosterosim, & tranquilizes the mind
Modified Blood Pressure Reducing Decoction	Reduces blood pressure, dreaminess, insomnia, brain fog, chest oppression
Modified Qi Replenishing and Yin Nourishing Formula	Stablizes blood pressure while recovering from dryness, exhaustion and headache
Modified Cardiotonic Formula	Reduces blood pressure, nourishes blood
Diabetes, kidney failure, aldosteronism, renovascular disease, coarctation of aorta, thyroid disease or tumors	Reduces blood pressure in serious cases, and coordinates with pharmaceutical drugs

11.2 PART 2: ANGINA PECTORIS

11.2.1 Pharmaceutical Drugs

Medicine	Action
Nitroglycerine	Vasodilator
Beta blocker	Blocks the binding receptor on heart, kidney, arteries, smooth muscle cells that respond to epinephrine, the stress response.
Calcium channel blocker	Antihypertensive that decreases blood pressure
ACE inhibitor	Vasodilator that lowers blood pressure
Antiplatelet	Inhibits formation of thrombus
Anticoagulant	Inhibits formation of thrombus
Aspirin	Antiplatelet drug

11.2.2 Chinese Medicine Formulas

Medicine	Action
Xu ming decoction	Clears heat and activates blood flow
Xue fu zhu yu tang	Moves blood, promotes circulation, stops pain
Tian ma gou teng yin	Calms liver, clears heat, tranquilizes the mind
Gui zhi tang	Relieves the exterior, harmonizes wei qi and ying qi
Pulse activating powder	Replenishes qi, nourishes and astringes yin
Dan shen dripping pills	Promotes blood circulation, removes stasis, relieves pain

11.3 PART 3: HYPERLIPIDEMIA

11.3.1 Pharmaceutical Drugs

Medicine	Action
Ezetimibe	Lowers cholesterol absorption in the small intestine.
Statins	Inhibits HMG-CoA reductase enzyme for cholesterol reduction.
Bile acid sequestrants	Prevents fat absorption in the intestines.
Niacin	Helps break down fat tissue.
Fibrate	Combines with statins to treat high cholesterol.

11.3.2 Chinese Medicine Formulas

Medicine	Action
Fleece flower formula	Reduces liver hyperactivity, relieves body distention, nourishes kidney yin and blood.
Weicao tang	Reduces blood lipids and increases lipid metabolism.
Gentian Formula	Purges liver fire to reduce adverse effects on the heart, pathogenic dampness transforming into heat.
Eliminate Blood Lipid Formula	Reduces blood lipids for weight loss.
Shen qi wan	Helps to tonify and strengthen the kidneys
Reduce Blood Lipid Formula	Reduces LDL levels.
Ginko leaf formula	Assists with the symptoms of lipoproteinemia, fights free radicals, moves blood and breaks up phlegm in the vessels.

11.4 PART 4: THROMBOANGIITIS OBLITERANS

11.4.1 Pharmaceutical Drugs

Medicine	Action
Hyperbaric oxygen therapy	Vasodilation
Prostaglandins	Vasodilation
Streptokinase	Antithrombosis
Corticosteroids	Anti -inflammatory

11.4.2 Chinese Medicine Formulas

Medicine	Action
Tao hong si wu tang	Breaks blood stasis, moves blood
Si miao yong an tang	Treats gangrene
Wu wei xiao du yin	Clears toxic heat and toxins
Xian fang huo ming yin	Clears toxic heat and toxins
Tou nong san	Removes toxins, replenishes qi and blood
Yang he tang	Treats gangrene

11.5 PART 5: TAKAYASU ARTERITIS

11.5.1 Pharmaceutical Drugs

Medicine	Action
Corticosteroids	Help control symptoms during the inflammatory process.
Cytotoxic agents	Immuno-suppressant medication combined with steroids for advanced stages when relapses are more frequent.
TNF inhibitors	Combine with corticosteroids and cytotoxic agents during remission stages to decrease immune responses.
Antiplatelets	Inhibit platelet aggregation causing thrombosis, and renal failure.
Anticoagulants	Reduce or blocks further thrombosis.
Calcium channel blocker	Treats hypertension.

11.5.2 Chinese Medicine Formulas

Medicine	Action
Si miao yong an tang	Clears heat, detoxifies and activates blood.
Wen yang tong mai tang	Activates blood, relieves pain, invigorates deficiency.
Tong mai huo xue tang	Activates blood and removes stasis, regulates flow of qi, replenish blood.
Bu yang huan wu tang	Invigorates qi, resolves blood stasis.
Bu pian yu feng tang	Replenishes qi, soothes liver and stops wind.

11.6 PART 6: BASIC METHOD FOR INDICATING OR CONTRA-INDICATING

- Patients in the moderate hypertension range where the blood pressure seems to fluctuate between 140–150 and 90–100 may benefit more with combining Chinese and Western medicine. When the blood pressure is elevated, specific symptoms may include anxious feelings in the chest, tension in the shoulders, heightened irritability, and either one of these additional symptoms: insomnia in the form of racing thoughts at night; or exhaustion during the day and inability to sleep at night, or exhaustion at various times during the day and night with more wakeful energy during the middle of the night.
- The use of Chinese herbal medicines for patients with angina pectoris requires timing. Prevention and control of symptoms during the uncertain diagnoses and stable angina stage is necessary to avoid future complications.
- Chinese medicine formulas are as effective as pharmaceutical drugs for lowering blood lipids. In some cases where a patient has been taking drugs to lower lipids and is indicated for Chinese medicine, the patient can be weaned off of the drugs and started on Chinese medicine within a few days.
- As in Western medicine, treatments using Chinese medicine should be seen as a palliative care option which mostly addresses symptoms. Chinese medicine practitioners and Western medicine practitioners should collaborate with the patient to document medicines and methods which have helped to make the most progress and prolong remission periods.

- During active or stable phases, it is suggested that herbal formulas be contraindicated if anticoagulant medications are used, because they provide equal pharmacodynamic action. If the formulas are used, they must be deeply modified for compatibility with the treatment course.

NOTES

Module Review Questions

1. Explain certain Western medicine key points about hypertension.
2. Explain certain Western medicine key points about rheumatic fever.
3. Explain certain Western medicine key points about endocarditis.
4. Discuss the similarities.
5. Discuss the differences.
6. Explain certain Chinese medicine key points about hypertension.
7. Explain certain Chinese medicine key points about rheumatic fever.
8. Explain certain Chinese medicine key points about endocarditis.
9. Discuss the similarities.
10. Discuss the differences.
11. List the pharmaceutical drugs and actions for each disorder.
12. List the basic herbal formulas and action for each disorder.
13. Discuss the treatment suggestions.

Log on at www.niambiwellness.com to access the companion course and quiz for Module 4.

SECTION V

Genetic Disorders in Cardiology

CHAPTER 12

Genetic Disorders in Cardiology

CHAPTER OBJECTIVES

After studying this chapter, you should be able to:

1. List the genetic disorders found in cardiac rhythm diseases.
2. List the genetic disorders found in chamber and valve diseases.
3. List the genetic disorders found in vascular system diseases.

12.1 PART 1: CARDIAC RHYTHM DISEASES

12.1.1 Arrhythmia

12.1.1.1 MicroRNA

MicroRNA (miRNA) regulates the expression of genes throughout the body including the cardiovascular system. Cardiovascular diseases where miRNA has contributed include valvular disease and atrial fibrillation.

12.1.1.2 ACE Gene Polymorphism

Patients with this gene may experience altered responses to ACE inhibitor medication.

Advanced Clinical Therapies in Cardiovascular Chinese Medicine. DOI: http://dx.doi.org/10.1016/B978-0-12-800122-6.00012-7

12.1.1.3 KCN (Potassium Channel) Gene Mutations and Familial Atrial Fibrillation

KCNE2	Romano-Ward syndrome/ long QT syndrome
KCNJ2	Short QT syndrome Andersen-Tawil syndrome
KCNQ1	Jervell and Lange-Nielsen syndrome Romano-Ward syndrome/ long QT syndrome (LQTS) Short QT syndrome Long QT syndrome
(FAF)	Romano-Ward syndrome/ long QT syndrome Short QT syndrome Andersen-Tawil syndrome Jervell and Lange-Nielsen syndrome Long QT syndrome

12.1.2 Sick Sinus Syndrome

12.1.2.1 *HCN4* Mutations

HCN channels are involved in the automaticity of the SA node, and mutations may be involved with the abrupt switch to sinus bradycardia from tachycardia, which is responsible for reduced excitability and characteristic in sick sinus syndrome.

12.2 PART 2: CHAMBER AND VALVE DISEASES

12.2.1 Hypertension

- Rennin–angiotensin system and sodium handling genes are associated with determining blood pressure.

12.2.1.1 Rare Childhood Disorders

Bartter syndrome	Hypertension symptoms present in infants and children.
Gitelman syndrome	Symptoms present in children or later in life in adults.
Liddle syndrome	Hypertension is autosomal dominant with hypertension which often begins in childhood featuring hypokalemia and aldosterone deficiency.
Pseudohypoaldosteronism type 1	Symptoms are seen in newborn infants and in failure to thrive children

12.2.2 Rheumatic Fever

12.2.2.1 The M Protein

M protein serotypes of streptococcus cell walls may be similar to the antigen located on myocardial cells and therefore such antibodies may attack cardiac myocytes.

12.2.2.2 *HLA-DR7* (Human Leukocyte Antigen, HLA)

This protein is a major histocompatibility complex (MHC) class II molecule which has a role of increasing the possibility of autoimmune reactions in some patients. It may be connected to CD4+ responses.

12.2.3 Endocarditis

12.2.3.1 *HLA-DQA1*

This gene is associated with autoimmunity and idiopathic membranous nephropathy, which causes nephrotic syndrome and leads to end-stage renal disease.

12.2.3.2 *CFHR1* and *CFH3*

These genes which are part of the MHC are involved with IgA nephropathy, which is connected to diseases with autoimmune and inflammatory complications such as in multiple sclerosis and systemic lupus.

12.3 PART 3: VASCULAR SYSTEM DISEASES

12.3.1 Hypertension

Mendelian forms of hypertension and hypotension are directly related to renal sodium reabsorption.

12.3.1.1 *MYH9*

This rare gene is related to nephritic syndrome where the blood pressure becomes a critical concern leading to kidney failure.

12.3.2 Angina Pectoris

12.3.2.1 *DKN2A* and *DKN2B*

These genes are involved with regulating cellular aging and lysis and the formation of plaque within the arteries.

12.3.2.2 *MTAP*

High homocysteine level is a risk factor for atherosclerosis. This gene processes cellular waste products into methionine which metabolizes excess homocysteine. Mutations are involved with failures in these processes.

12.3.3 Lipoprotein Disorders

12.3.3.1 Familial Hypercholesterolemia

Type I Familial hyperchylomicronemia
-Lipoprotein lipase (LPL) deficiencies -Elevations in triglycerides -Reduced HDL levels.
Type II Familial hypercholesterolemia
-PCSK9 mutations with elevated cholesterol levels -LDL-R gene defect with elevated LDL levels -Apo B gene defects and high elevations of LDL -Linked to coronary artery disease -Prevalence in the U.S. African Americans
Type III Familial hyperlipoproteinemia
-Apo B gene defects -PCSK9 mutations, -LDLRAP1 mutations -ABCG 8 mutations -Excess liver VLDL makes excess cholesterol -Xanthomas -High cholesterol -Metabolic syndrome
Type V Familial hyperlipoproteinemia
-Combination of type I and IV -Splenomegaly -Xanthomas

Apo B100	-Elevated levels due to coronary artery disease due to *Staphyococcus* aureus, leaving cholesterol streaks.
Apo B48	-Excess in chylomicrons raises LDL levels and complicates Diabetes Mellitus treatment.
Apolipoprotein E (Apo E)	E2: slow fat metabolism, increased in risk of coronary artery disease
	E3: neutral allele
	E4: atherosclerosis, sleep apnea, ischemic cerebrovascular disease

12.3.3.2 Thromboangiitis Obliterans

12.3.3.2.1 Human Leukocyte Antigen

This includes the locus of immune system genes related to the MHC and are divided into three classes.

Class I	Related to digestion
Class II	Related to antigen formation
Class III	Related to the complement system or the innate system

12.3.3.3 *MS4A1*

CD3 and CD20: Deficiencies of both are related to complications associated with T and B lymphocytes. T-lymphocyte mediation leads to autoimmunity. Humoral immunity hyperfunction by B lymphocytes leads to vascular injury and thrombus formation.

NOTES

Course Review Questions

1. List the cardiovascular diseases which are commonly treated in the TCM clinic.
2. Describe the aims of treatment using TCM alone or integrating it with Western medicine.
3. Compare and contrast constitutional factors versus genetic determinants.
4. Explain the common TCM patterns and differentiations between different Western medicine diseases.
5. Describe the possible similarities between cardiovascular diseases with similar differentiations.
6. Describe the possible differences between cardiovascular diseases with similar differentiations.
7. Explain certain Western medicine key points about arrhythmia.
8. Explain certain Western medicine key points about sick sinus syndrome.
9. Explain certain Chinese medicine key points about arrhythmia.
10. Explain certain Chinese medicine key points about sick sinus syndrome.
11. List the pharmaceutical drugs and actions for each disorder.
12. Explain certain Western medicine key points about hypertension.
13. Explain certain Western medicine key points about rheumatic fever.
14. Explain certain Western medicine key points about endocarditis.
15. Explain certain Chinese medicine key points about hypertension.
16. Explain certain Chinese medicine key points about rheumatic fever.
17. Explain certain Chinese medicine key points about endocarditis.
18. List the pharmaceutical drugs and actions for each disorder.
19. Explain certain Western medicine key points about hypertension.
20. Explain certain Western medicine key points about rheumatic fever.
21. Explain certain Western medicine key points about endocarditis.
22. Explain certain Chinese medicine key points about hypertension.
23. Explain certain Chinese medicine key points about rheumatic fever.
24. Explain certain Chinese medicine key points about endocarditis.

This also concludes the Integrative Advanced Clinical Therapies in Cardiovascular Chinese Medicine course. It is strongly suggested that you log onto the courses at the companion websites to review the course modules. Next, submit course documents and complete the final exam.

Upon passing the exam, you will receive completion certificates which include your name and practice license number, along with the specific number of credit hours awarded for this course. Electronic transmission of CEU and PDA credits will be sent to NCCAOM and your state medical board.

'nd by CPI Group (UK) Ltd, Croydon, CR0 4YY
03/10/2024
01040427-0012